SCIENCE TOOLS

6 USING INVISIBLE LIGHT

John O.E. Clark

GROLIER
EDUCATIONAL

About this set

SCIENCE TOOLS deals with the instruments and methods that scientists use to measure and record their observations. Theoretical scientists apply their minds to explaining a whole range of natural phenomena. Often the only way of testing these theories is through practical scientific experiment and measurement—which are achieved using a wide selection of scientific tools. To explain the principles and practice of scientific measurement, the nine volumes in this set are organized as follows:

Volume 1—Length and Distance; Volume 2—Measuring Time; Volume 3—Force and Pressure; Volume 4—Electrical Measurement; Volume 5—Using Visible Light; Volume 6—Using Invisible Light; Volume 7—Using Sound; Volume 8—Scientific Analysis; Volume 9—Scientific Classification.

The topics within each volume are presented as self-contained sections, so that your knowledge of the subject increases in logical stages. Each section is illustrated with color photographs, and there are diagrams to explain the workings of the science tools being described. Many sections also contain short biographies of the scientists who discovered the principles that the tools employ.

Pages at the end of each book include a glossary that gives the meanings of scientific terms used, a list of other sources of reference (books and websites), and an index to all the volumes in the set. There are cross-references within volumes and from volume to volume at the bottom of the pages to link topics for a fuller understanding.

Published 2003 by Grolier Educational, Danbury, CT 06816

This edition published exclusively for the school and library market

Planned and produced by Andromeda Oxford Limited,
11-13 The Vineyard,
Abingdon, Oxon OX14 3PX

Copyright © Andromeda Oxford Limited

Project Director Graham Bateman
Editors John Woodruff, Shaun Barrington
Editorial assistant Marian Dreier
Picture manager Claire Turner
Production Clive Sparling

Design and origination by Gecko

Printed in Hong Kong

Library of Congress Cataloging-in-Publication Data

Clark, John Owens Edward.
  Under the microscope : science tools / John O.E. Clark.
    p. cm.
Summary: Describes the fundamental units and measuring devices that scientists use to bring systematic order to the world around them.
Contents: v. 1. Length and distance -- v. 2. Measuring time -- v. 3. Force and pressure -- v. 4. Electrical measurement -- v. 5. Using visible light -- v. 6. Using invisible light -- v. 7. Using sound -- v. 8. Scientific analysis -- v. 9. Scientific classification.
  ISBN 0-7172-5628-6 (set : alk. paper) -- ISBN 0-7172-5629-4 (v. 1 : alk. paper) -- ISBN 0-7172-5630-8 (v. 2 : alk. paper) -- ISBN 0-7172-5631-6 (v. 3 : alk. paper) -- ISBN 0-7172-5632-4 (v. 4 : alk. paper) -- ISBN 0-7172-5633-2 (v. 5 : alk. paper) -- ISBN 0-7172-5634-0 (v. 6 : alk. paper) -- ISBN 0-7172-5635-9 (v. 7 : alk. paper) -- ISBN 0-7172-5636-7 (v. 8 : alk. paper) -- ISBN 0-7172-5637-5 (v. 9 : alk. paper)
  1. Weights and measures--Juvenile literature. 2. Measuring instruments--Juvenile literature. 3. Scientific apparatus and instruments--Juvenile literature. [1. Weights and measures. 2. Measuring instruments. 3. Scientific apparatus and instruments.] I. Title: Science tools. II. Title.
  QC90.6 .C57 2002
  530.8--dc21

2002002598

About this volume

Volume 6 of *Science Tools* concerns itself with invisible radiation. That is the whole of the electromagnetic spectrum except for visible light, which is the only radiation that humans can see. "Invisible light" ranges from penetrating gamma rays and x-rays to microwaves and radio waves. More like visible light are ultraviolet light and infrared radiation, which we can feel as heat. Microwaves and radio waves coming from outer space are detected on Earth by radio telescopes. Beams of electrons can be made to behave like light rays—for example, they can be focused by electromagnetic lenses, as they are in the various kinds of electron microscopes.

Contents

Main units of measurement

Scientists spend much of their time looking at things and making measurements. These observations allow them to develop theories, from which they can sometimes formulate laws. For example, by observing objects as they fell to the ground, the English scientist Isaac Newton developed the law of gravity.

To make measurements, scientists use various kinds of apparatus, which we are calling "science tools." They also need a system of units in which to measure things. Sometimes the units are the same as those we use every day. For instance, they measure time using hours, minutes, and seconds—the same units we use to time a race or bake a cake. More often, though, scientists use special units rather than everyday ones. That is so that all scientists throughout the world can employ exactly the same units. (When they don't, the results can be very costly. Confusion over units once made NASA scientists lose all contact with a space probe to Mars.) A meter is the same length everywhere. But everyday units sometimes vary from country to country. A gallon in the United States, for example, is not the same as the gallon people use in Great Britain (a U.S. gallon is about one-fifth smaller than a UK gallon).

On these two pages, which for convenience are repeated in each volume of *Science Tools*, are set out the main scientific units and some of their everyday equivalents. The first and in some ways most important group are the SI units (SI stands for Système International, or International System). There are seven base units, plus two for measuring angles (Table 1). Then there are 18 other derived SI units that have special names. Table 2 lists the 11 commonest ones, all named after famous scientists. The 18 derived units are defined in terms of the 9 base units. For example, the unit of force (the newton) can be defined in terms of mass and acceleration (which itself is measured in units of distance and time).

▼ Table 1. **Base units** of the SI system

QUANTITY	NAME	SYMBOL
length	meter	m
mass	kilogram	kg
time	second	s
electric current	ampere	A
temperature	kelvin	K
luminous intensity	candela	cd
amount of substance	mole	mol
plane angle	radian	rad
solid angle	steradian	sr

▼ Table 2. **Derived SI units** with special names

QUANTITY	NAME	SYMBOL
energy	joule	J
force	newton	N
frequency	hertz	Hz
pressure	pascal	Pa
power	watt	W
electric charge	coulomb	C
potential difference	volt	V
resistance	ohm	Ω
capacitance	farad	F
conductance	siemens	S
inductance	henry	H

▼ **Table 3. Metric prefixes** for multiples and submultiples

PREFIX	SYMBOL	MULTIPLE
deka-	da	ten ($\times 10$)
hecto-	h	hundred ($\times 10^2$)
kilo-	k	thousand ($\times 10^3$)
mega-	M	million ($\times 10^6$)
giga-	G	billion ($\times 10^9$)

PREFIX	SYMBOL	SUBMULTIPLE
deci-	d	tenth ($\times 10^{-1}$)
centi-	c	hundredth ($\times 10^{-2}$)
milli-	m	thousandth ($\times 10^{-3}$)
micro-	µ	millionth ($\times 10^{-6}$)
nano-	n	billionth ($\times 10^{-9}$)

Scientists often want to measure a quantity that is much smaller or much bigger than the appropriate unit. A meter is not much good for expressing the thickness of a human hair or the distance to the Moon. So there are a number of prefixes that can be tacked onto the beginning of the unit's name. The prefix milli-, for example, stands for one-thousandth. Therefore a millimeter is one-thousandth of a meter. Kilo- stands for one thousand times, so a kilometer is 1,000 meters. The commonest prefixes are listed in Table 3.

Table 4 shows you how to convert from everyday units (known as customary units) into metric units, for example from inches to centimeters or miles to kilometers. Sometimes you may want to convert the other way, from metric to customary. To do this, divide by the factor in Table 4 (not multiply). So, to convert from inches to centimeters, *multiply* by 2.54. To convert from centimeters to inches, *divide* by 2.54. More detailed listings of different types of units and their conversions are given on pages 6–7 of each volume. You do not have to remember all the names: They are described or defined as you need to know them throughout *Science Tools*.

▶ **Table 4. Conversion** to metric units

TO CONVERT FROM	TO	MULTIPLY BY
inches (in.)	centimeters (cm)	2.54
feet (ft)	centimeters (cm)	30.5
feet (ft)	meters (m)	0.305
yards (yd)	meters (m)	0.914
miles (mi)	kilometers (km)	1.61
square inches (sq in.)	square centimeters (sq cm)	6.45
square feet (sq ft)	square meters (sq m)	0.0930
square yards (sq yd)	square meters (sq m)	0.836
acres (A)	hectares (ha)	0.405
square miles (sq mi)	hectares (ha)	259
square miles (sq mi)	square kilometers (sq km)	2.59
cubic inches (cu in.)	cubic centimeters (cc)	16.4
cubic feet (cu ft)	cubic meters (cu m)	0.0283
cubic yards (cu yd)	cubic meters (cu m)	0.765
gills (gi)	cubic centimeters (cc)	118
pints (pt)	liters (l)	0.473
quarts (qt)	liters (l)	0.946
gallons (gal)	liters (l)	3.79
drams (dr)	grams (g)	1.77
ounces (oz)	grams (g)	28.3
pounds (lb)	kilograms (kg)	0.454
hundredweights (cwt)	kilograms (kg)	45.4
tons (short)	tonnes (t)	0.907

The electromagnetic spectrum

Light is a form of radiation. In fact, it is just a small part of a whole range of radiation, all of a similar type. The range goes from short-wavelength gamma rays and x-rays to long-wavelength microwaves and radio waves. All these rays and waves have many uses in science.

Because we can see it, light is known to scientists as visible radiation. But light is only part of a wide range of other radiations that we cannot see—think of them as kinds of invisible light. And just as the range of colors in visible light is called a spectrum, the range of these invisible radiations is also a spectrum, known as the electromagnetic spectrum.

Wavelengths shorter than light

Volume 5 of *Science Tools* explains how the different colors in the visible spectrum have different wavelengths. In a similar way the radiations on different parts of the electromagnetic spectrum also have different wavelengths. Gamma rays, a type of radiation that is given off by many radioactive substances, have the shortest waves:

◄ **To human eyes** this daisy is bright yellow. The picture was taken in ultraviolet light and shows how the flower might look to an insect such as a bee. The lines that show up on the petals are called nectar guides because they "steer" the bee toward the center of the flower, where the nectar is located.

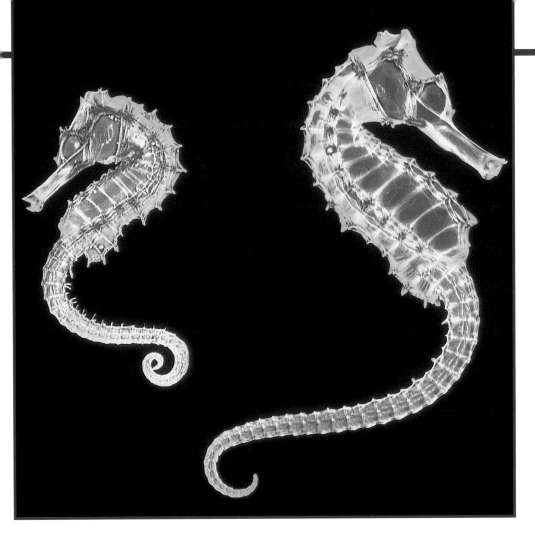

◀ **See-through sea horses** photographed in x-rays. These strange fish live in tropical and subtropical seas, where they cling to plants with their long coiled tails.

Their wavelength can be as short as 10^{-14} meters—or, in "longhand," 0.00000000000001 meters. That is ten thousand times smaller than the smallest atom! This enables gamma rays to penetrate materials easily, making them dangerous to living things—but it also means that they are very useful to scientists.

Next along the wavelength scale of the electromagnetic spectrum come x-rays, which have wavelengths of between 10^{-8} and 10^{-10} meters. There is no clear-cut division between the two types of radiation, and the shortest x-rays blend into the longest gamma rays. X-rays are also short enough to penetrate materials, which gives them many important uses in science, medicine, and industry. (We look more closely at gamma rays and x-rays on pages 26–31.)

At longer wavelengths still, x-rays give way to ultraviolet radiation, also known as ultraviolet light (*ultra* means "beyond," and ultraviolet light is just beyond the violet end of the visible spectrum). It has a wavelength in the region of 10^{-6} meters. Human beings cannot see ultraviolet radiation, but it is important to us. It forms part of sunlight, and our bodies use it to make vitamin D, which helps build strong, healthy bones. Deprived children who do not get enough ultraviolet from sunlight can develop a disease called rickets that deforms their bones, especially if there is not enough calcium in their diet. Ultraviolet light from the Sun also gives us a suntan. But too much of it can be harmful, and it can cause skin cancer.

Although we cannot see ultraviolet light, some insects such as bees and butterflies can see it. Pigments that color some flowers can be seen only in ultraviolet light or by creatures with ultraviolet vision. Often the pigments form lines or patterns that can, for example, direct a bee toward the center of a flower, where it can find the pollen and nectar. (Scientific uses of ultraviolet light are described on pages 10–13.)

The electromagnetic spectrum

Electromagnetic spectrum

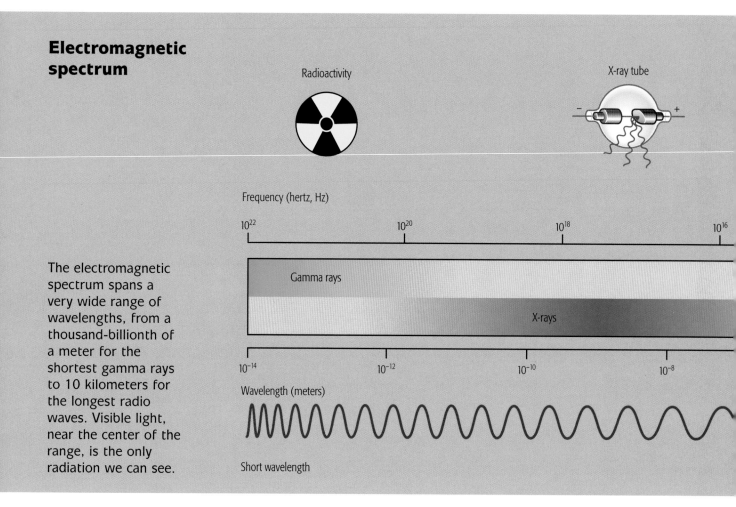

Radioactivity

X-ray tube

The electromagnetic spectrum spans a very wide range of wavelengths, from a thousand-billionth of a meter for the shortest gamma rays to 10 kilometers for the longest radio waves. Visible light, near the center of the range, is the only radiation we can see.

Frequency (hertz, Hz)

10^{22} 10^{20} 10^{18} 10^{16}

Gamma rays

X-rays

10^{-14} 10^{-12} 10^{-10} 10^{-8}

Wavelength (meters)

Short wavelength

Wavelengths longer than light

Just past the red end of the visible spectrum, with a wavelength of about 10^{-4} meters, or 0.1 millimeters, is the region of infrared radiation. *Infrared* means "below red." We experience this type of electromagnetic radiation as heat rays— the warmth of the Sun that we feel on our skin on a sunny day is infrared radiation.

Like ultraviolet, infrared is invisible to human eyes. But some animals have their own infrared sensors. One of them is the pit viper, a snake that can detect its warm-blooded prey in the pitch-black of the desert night by the infrared radiation given off by its victim's body. Scientists have developed similar infrared detectors. Police helicopters use them in pursuit of suspects at

night, and rescuers use them to search for people trapped in buildings after an earthquake. (For other applications of infrared radiation see pages 14–19.)

Next in line along the electromagnetic spectrum come microwaves. They are high-frequency radio waves with wavelengths of around 10^{-2} meters, or 1 centimeter. They have many uses, including microwave ovens, radar, and satellite communications (see pages 32–35).

The radio waves that are used for everyday communications span the wavelength range from 1 to 10^4 meters (1 meter to 10 kilometers). They are often divided into shortwave (as used for round-the-world broadcasts), medium-wave (for ordinary AM broadcasts), and long-wave

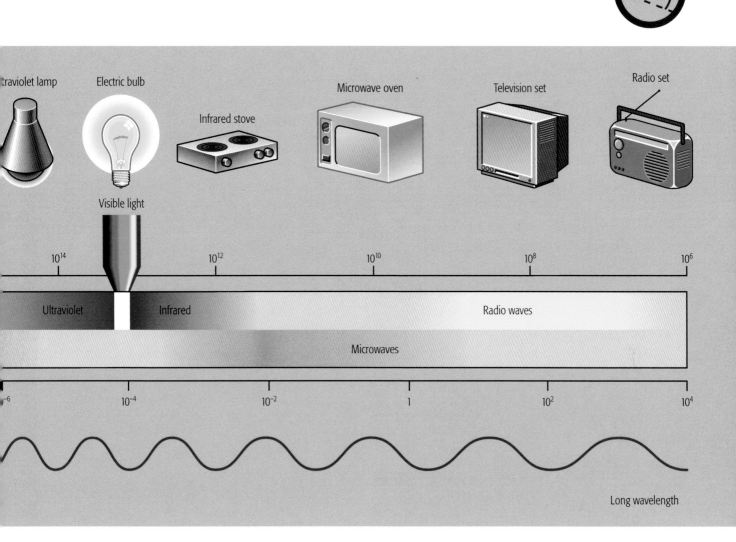

Ultraviolet lamp
Electric bulb
Infrared stove
Microwave oven
Television set
Radio set
Visible light

10^{14} 10^{12} 10^{10} 10^8 10^6

Ultraviolet   Infrared   Radio waves

Microwaves

10^{-6} 10^{-4} 10^{-2} 1 10^2 10^4

Long wavelength

(anything over 1,000 meters). Some stars and other celestial objects give off radio waves, which are studied by radio astronomers (as described on pages 36–39).

Speed and frequency

All types of electromagnetic radiation travel at the same speed—the speed of light. It is the fastest speed known—in fact, the fastest possible speed—and equals nearly 300,000 kilometers per *second* (about 186,000 miles per second). That is for radiation traveling in a vacuum. The speed is very slightly slower for radiation that is traveling through a medium such as air or water, but the difference is so small that it can be ignored for most purposes.

Electromagnetic radiation is a type of wave motion—a succession of waves. The number of waves passing a given point in 1 second is known as the frequency of the radiation. Frequency is measured in a unit called the hertz, abbreviated to Hz. A frequency of 1 Hz represents one complete wave per second. As you can see from the illustration above, the range of frequency and wavelength across the whole electromagnetic spectrum is enormous. The frequencies range from about 10^{22} (10 billion trillion) hertz for the shortest-wavelength gamma rays to as little as 10^5 (100,000) hertz for the longest radio waves. So you can see that the wavelengths and frequencies of electromagnetic radiation involve some of the smallest and some of the largest numbers in science.

Ultraviolet radiation

Ultraviolet radiation lies just outside the visible spectrum, on the short-wavelength side. It reaches the Earth from the Sun. It can be harmful to life, but most of it is blocked by the ozone layer in the atmosphere before it can get to the ground. Ultraviolet radiation can also be useful.

Ultraviolet radiation—usually abbreviated to UV—is not the only radiation to reach the Earth from space. The illustration on the opposite page shows a cross-section of the Earth's atmosphere. The full electromagnetic spectrum is pictured at the top in order of wavelength, the shortest on the left and the longest on the right.

We can see at once how lucky we are to have an atmosphere, because it blocks most of the harmful radiation before it can reach the ground. The planet Mercury, closest to the Sun, has almost no atmosphere and is constantly bathed in the whole range of radiations. Gamma rays, the most harmful of all, are stopped by the time they reach the Earth's stratosphere, about 20 km (roughly 15 miles) above the surface. X-rays don't even get that far and are absorbed by the thin atmosphere 80–90 km (about 50–55 miles) up.

Most of the gamma rays and x-rays that arrive at the Earth come from the Sun. The Sun is also the main source of ultraviolet, most of which is absorbed by the layer of ozone in the atmosphere at an altitude of about 30 km (about 20 miles). But some UV does get through and can give us a suntan or even skin cancer. The harm that UV can cause is what makes scientists very concerned about the breakdown of the ozone layer. The ozone is being destroyed by atmospheric pollution, particularly by gases called CFCs that are used in iceboxes and as propellants in aerosol cans.

Visible light obviously penetrates the atmosphere completely—otherwise, down here at the surface it would be a very dark place, even in daytime! Some infrared radiation gets through, allowing the Sun's heat to reach the ground. Much of this infrared is reradiated from the ground at a slightly longer wavelength and can be absorbed by gases in the atmosphere.

The longest wavelengths, belonging to microwaves and radio waves, all easily penetrate the atmosphere. For this reason radio astronomers can build their giant radio telescopes practically anywhere on Earth. But astronomers who want to study the complete range of ultraviolet and infrared radiation from celestial objects have to place their instruments on artificial satellites that orbit above the Earth's obscuring atmosphere.

Producing ultraviolet light

Ultraviolet has several useful applications. As well as for the lamps in tanning beds, from which people can get an artificial tan, it is used for killing germs and insects and in making printed circuits and microchips. For these purposes it can be produced by a UV fluorescent tube, illustrated on page 12. It is similar to the ordinary fluorescent tube used in lighting for homes and offices, but it differs in one important way. It consists of a long glass tube from which all the air has been removed. A little mercury vapor is then added to

▶ **The Earth's atmosphere** blocks nearly all the radiation that comes from space. Only visible light and some ultraviolet and infrared penetrate all the way to the ground.

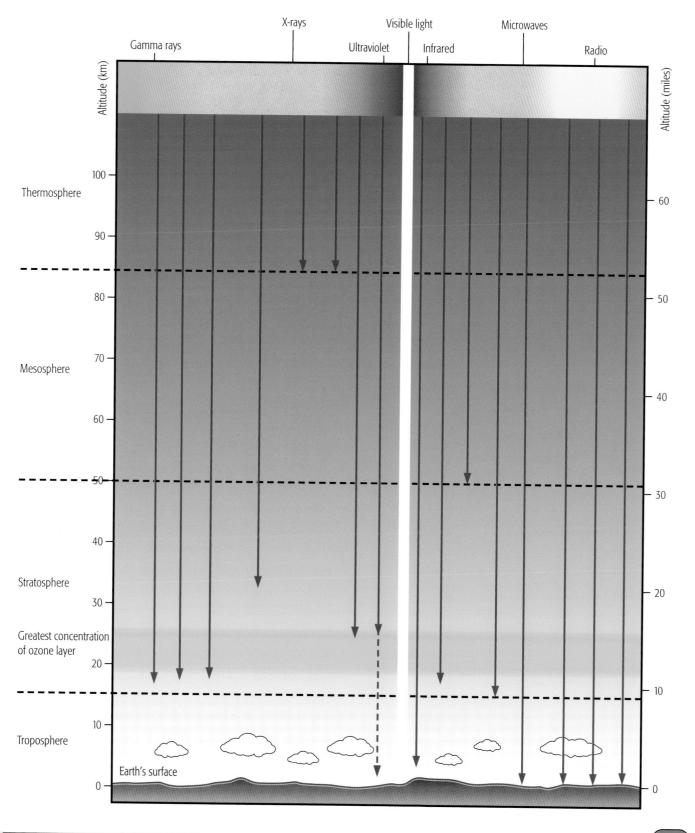

Gamma rays   X-rays   Ultraviolet   Visible light   Infrared   Microwaves   Radio

Altitude (km)

Thermosphere

100

90

80

70

Mesosphere

60

50

40

Stratosphere

30

Greatest concentration
of ozone layer

20

10

Troposphere

Earth's surface

0

Altitude (miles)

60

50

40

30

20

10

0

 FOR MORE ON ULTRAVIOLET RADIATION SEE *MAKING A SPECTRUM* **5:10**; *INFRARED RADIATION* **6:14**; *SPECTROSCOPES* **8:20**

11

Ultraviolet radiation

▼ **The output** from a UV fluorescent tube depends on the actions of electrons produced by a heated filament at one end of the tube.

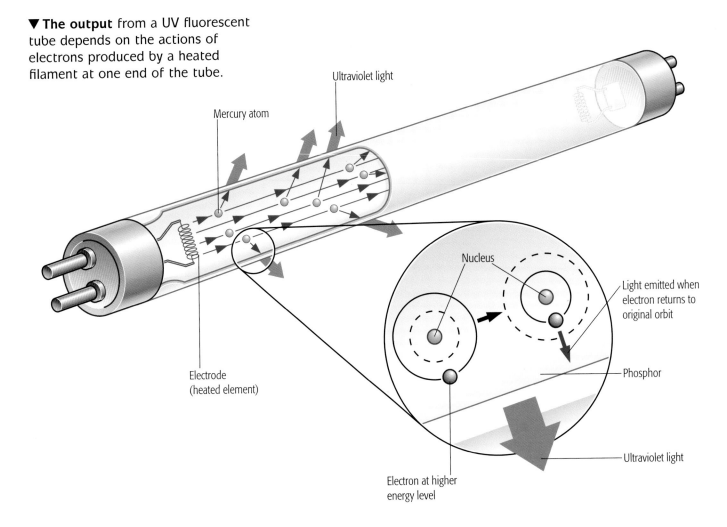

Ultraviolet light

Mercury atom

Electrode (heated element)

Nucleus

Light emitted when electron returns to original orbit

Phosphor

Ultraviolet light

Electron at higher energy level

the tube. There are electrodes at each end by which an electric current can enter and leave. One of the electrodes is in the form of a filament that gets red hot and emits a stream of electrons (negatively charged subatomic particles).

What happens next involves the mercury atoms in the tube. Each atom (of any element, not just mercury) consists of a central, positively charged nucleus with electrons in orbit around it—a bit like planets orbiting the Sun. Each orbit represents a different amount of energy, and physicists refer to these orbits as the energy levels of the electrons.

When an electron from the tube's hot filament hits a mercury atom, it passes on its energy to an orbiting electron, which it "knocks" into a higher

energy level. That makes the mercury atom unstable. It soon becomes stable again because the so-called excited electron rapidly returns to its original orbit. But to do so it has to get rid of its extra energy, which it does by giving off light— ordinary visible light.

On the inside of the glass tube, like a coat of paint, is a layer of a chemical substance called a phosphor. The particular phosphor in the UV tube gives off UV light when it is illuminated by ordinary light. That is what makes the UV tube different: The phosphor in an ordinary fluorescent tube gives off only visible light. The light emitted by the mercury atoms makes the phosphor give off UV, which shines out from the tube. There is also a little visible light emitted, and to human eyes the tube seems to have a rich, lilac glow.

The ultraviolet spectrum

Just as visible light has a spectrum, UV radiation spans a wide enough range of wavelengths to have a "minispectrum" of its own. Its wavelengths are usually expressed, like those of visible light, in nanometers (abbreviated to nm), where 1 nanometer = 10^{-9} meters (one-billionth of a meter).

One way of subdividing the part of the UV spectrum that reaches the Earth's surface is based on its potentially harmful effects on human skin. The longest wavelength range, 320–400 nm, is called UVA; the middle range, 290–320 nm, is UVB; and the shortest range, 230–290 nm, is UVC. Normal doses of UVA are not harmful to the skin and are used to treat skin disorders (such as the itching condition psoriasis). UVB is the form of UV that gives you a suntan. Many doctors think that the short-wavelength UVC can cause skin cancer.

Astronomers are concerned with a wider UV band and divide it into extreme UV (EUV, 10–100 nm), far UV (FUV, 100–200 nm), and near UV (NUV, 200–320 nm).

Ultraviolet astronomy

The first UV satellites were launched in the late 1960s. A major advance came with the International Ultraviolet Explorer (IUE), which carried a UV telescope with an aperture (diameter) of 45 centimeters (17.7 in.). Because glass absorbs UV radiation, the telescope's optics had to be made from transparent minerals such as quartz and fluorite. The IUE operated from 1978 to 1996, making it the longest-lived astronomical satellite.

The IUE was followed in 1990 by Rosat, a satellite that carried a wide-field UV camera and worked until 1998. The Extreme Ultraviolet

Explorer (EUVE), launched by NASA in 1992, made a UV survey of the whole sky. The Hubble Space Telescope (HST), placed into orbit by NASA in 1990, can also "see" in the UV region. Its instruments have reflectors made from the rare metal iridium, which unlike other materials does not absorb UV. Astronomers hope that the HST will go on working at least until 2010.

▲ **Traces of uranium** in this glass bowl make it fluoresce brightly when it is illuminated by ultraviolet radiation. The UV excites the uranium atoms, just as electrons excite mercury atoms in a fluorescent tube.

Infrared radiation

In the electromagnetic spectrum infrared radiation lies just past the long-wavelength end of the visible spectrum. Like ultraviolet, it also reaches the Earth from the Sun, and we can feel it as heat rays on a sunny day. All hot objects, not just the Sun, emit infrared radiation.

The wavelengths of infrared radiation are longer than those of visible light and shorter than those of microwaves, which are next to them in the electromagnetic spectrum. Infrared, or IR for short, is given off by all hot objects, and we can feel it as heat rays. It is given off by the Sun, by other stars, and by some types of nebulae (a nebula is a huge cloud of gas in space). The study of these objects is called infrared astronomy. Infrared radiation can also be used for heating things, and infrared lamps and ovens have many applications, from drying paint to cooking food.

The shortest-wavelength infrared radiation from the Sun passes right through the Earth's atmosphere (see the illustration on page 11). It also passes through any haze there might be in the air. Haze scatters ordinary light, so infrared cameras are used for taking pictures of distant scenes obscured by haze. Infrared photography was invented about 120 years ago and today is important in medical diagnosis, agriculture, and industry.

Medical infrared photographs are called thermograms. They can reveal abnormal growths such as tumors, which are warmer than the

surrounding tissue. Infrared images of electronic circuits can reveal defects where components are running too hot. In infrared photographs taken from airplanes or orbiting satellites, like those on these pages, vegetation has a distinct color. Crops that have been attacked by disease or insects stand out because damaged plants do not reflect infrared as strongly as healthy plants do. Aerial infrared photographs are also used by geologists searching for new deposits of minerals.

Other uses of infrared are in the focusing mechanism of autofocus cameras, optical devices for "seeing" in the dark, and security alarms that are triggered when an intruder breaks an invisible beam of infrared. It also has important applications in analytical chemistry, as described in detail in Volume 8 of *Science Tools*.

Detecting infrared radiation

William Herschel discovered infrared radiation in 1800 when he split sunlight into the colors of the rainbow (the solar spectrum) by shining it through a glass prism. When he held a thermometer just beyond the red end of the spectrum, he noticed a rise in temperature. This rise was caused by heat rays—infrared radiation—from the Sun.

A more sensitive instrument for detecting infrared (and other radiation) is called a bolometer. The first of them was made by the U.S. astronomer and inventor Samuel Pierpont Langley in 1880 for measuring radiation from stars. This type of bolometer makes use of the fact that the electrical resistance of a blackened strip of

◀ **This infrared photograph** of the countryside shows an expressway cutting through farmland. The vegetation appears red because the chlorophyll it contains reflects infrared radiation.

William Herschel

Frederick William Herschel was born in 1738 in Hanover, Germany. His father was a musician, and when he was fourteen, William joined a military band in Hanover. After he moved to England in 1757, he continued to work as a musician. But he soon took up astronomy, and with the help of his sister Caroline he made many major discoveries, becoming the most famous astronomer of his time.

In 1773, Herschel began to build his own telescopes. In 1781 he found Uranus, the first major planet to be discovered in modern times. Six years later he found two of its moons, and in 1789 he observed two new moons of Saturn. He also cataloged many double stars and nebulae. He was the first to suggest that double stars may be orbiting around each other.

In 1800 he was holding a thermometer just beyond the red end of the Sun's spectrum when he noticed a rise in temperature—he had discovered invisible infrared radiation. Although Herschel was an astronomer, he could not have predicted that from this discovery would develop one of the most important branches of modern astronomy. He died in 1822.

His son John, who had collaborated with him on many observing programs, became one of the greatest British scientists of the 19th century.

FOR MORE ON INFRARED RADIATION SEE *MAKING A SPECTRUM* 5:10; *ULTRAVIOLET RADIATION* 6:10; *SPECTROSCOPES* 8:20

platinum metal changes when it is heated by infrared radiation. In a modern semiconductor bolometer the platinum is replaced by a photoconductor such as silicon or lead sulfide. The photoconductor is a substance whose electrical resistance falls when it is exposed to infrared radiation, and this fall can be measured.

Another type of device, called a thermocouple, can be used as an infrared detector. It too makes use of electrical effects. Two wires of different metals, or two rods of different semiconductors, are joined at their ends to form a circuit. One join, called the cold junction, is kept cool, often by being dipped in liquid air. The infrared radiation heats the other join, called the hot junction. Because the two junctions are at different temperatures, a voltage flows in the circuit. The size of the voltage depends on the temperature difference.

Objects giving off infrared radiation or reflecting it can be viewed—and photographed—through an image intensifier. A lens gathers infrared radiation, together with what little light there is, and focuses it onto a charge-coupled device (CCD), as used in a modern camcorder. This produces a pattern of

electrons that go on to strike more CCDs. The much larger electron output from these CCD "amplifiers" creates a greenish image on a fluorescent screen (see the illustration below).

Producing infrared radiation

Another source of infrared is an infrared laser. *Laser* stands for **l**ight **a**mplification by **s**timulated **e**mission of **r**adiation, and here the "light" is infrared radiation. The laser action is produced in a gas such as carbon dioxide or carbon monoxide. Carbon dioxide infrared lasers are used in light radar (LIDAR) systems, which work with infrared rays instead of microwave beams, and in apparatus used to separate isotopes (different forms of the same chemical element). Infrared can also be produced by LEDs (light-emitting diodes) based on a semiconductor such as gallium arsenide. They are used in optical fibers for computers and in communications systems such as cable TV.

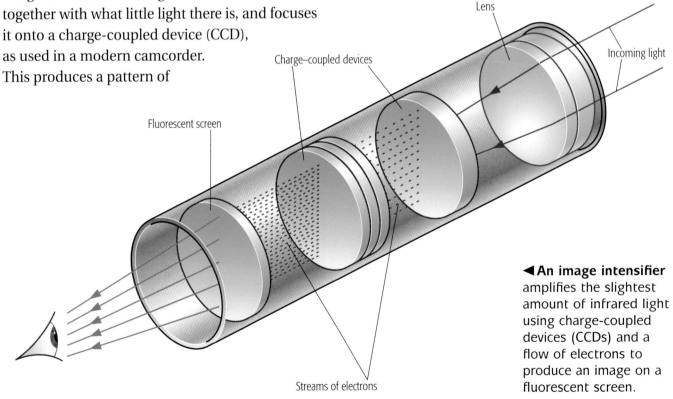

Lens

Incoming light

Charge–coupled devices

Fluorescent screen

Streams of electrons

◀**An image intensifier** amplifies the slightest amount of infrared light using charge-coupled devices (CCDs) and a flow of electrons to produce an image on a fluorescent screen.

▲ **This aerial photograph** of the Isle of Wight, off the
south coast of England, was taken by the French SPOT-1
infrared satellite. The brown areas are woodland, while
water shows up blue-black because it absorbs infrared
radiation. The pale blue patches are buildings.

Infrared radiation

The infrared spectrum

The part of the electromagnetic spectrum between visible light and microwaves—the infrared region—forms a "minispectrum" of its own. Its wavelength range runs from about 1 micrometer (1 μm) to 1,000 μm. It is subdivided as follows: 0.8 to 8 μm is called the near infrared, 8 to 30 μm is termed the mid-infrared, and 30 to 300 μm is the far infrared. Radiation that has wavelengths longer than 300 μm, grading into microwaves, is known as submillimeter radiation. Only the shorter infrared wavelengths penetrate the Earth's atmosphere. Longer wavelengths are absorbed by water vapor and other gases in the air.

Infrared Earth satellites

The first artificial satellites to make use of infrared were weather satellites called TIROS (Television and Infra Red Observation Satellite). Beginning in 1960, NASA launched ten of these satellites to monitor heat radiated from the Earth into space and to photograph global cloud cover to aid in tracking storms and in weather forecasting. In all, they sent back about 650,000 pictures.

Since then, many other Earth satellites have carried infrared cameras as part of their equipment. From 1966 to 1969 NASA launched nine TOS (Tiros Operational System) satellites, followed in 1970–76 by six ITOS (Improved TOS) satellites. From 1972 the Landsat series of satellites surveyed the Earth at infrared wavelengths. The infrared picture on page 17 is one of many taken by the French SPOT-1 satellite.

▶ The Infrared Astronomical Satellite (IRAS) was launched into Earth orbit by NASA to study sources of infrared radiation. It carried a reflecting telescope and infrared detectors.

Infrared astronomy

Because much of the infrared radiation from space is blocked by water vapor in the Earth's atmosphere, infrared observations of celestial objects have to be made from high-altitude observatories or from satellites in orbit around the Earth. Balloons, rockets, and high-flying airplanes have also been used for infrared astronomy. Targets for infrared telescopes include the atmospheres of the planets, the clouds of gas and dust (nebulae) that lie between the stars, red giant stars, and distant galaxies. From infrared observations of our own galaxy, the Milky Way, astronomers think that there is probably a massive black hole at its center.

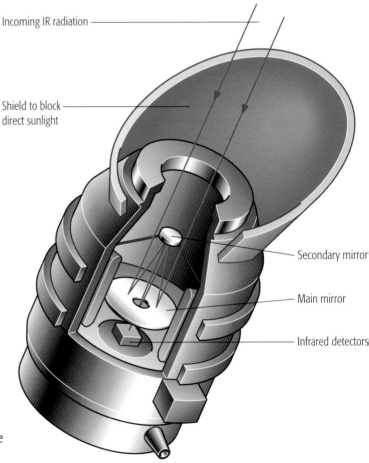

Incoming IR radiation

Shield to block direct sunlight

Secondary mirror

Main mirror

Infrared detectors

Infrared astronomy uses reflecting telescopes based on mirrors, not lenses (as described in Volume 5 of *Science Tools*). One of the most important sites for ground-based observatories is the 4,200-meter (13,800-ft) extinct volcano Mauna Kea on the island of Hawaii. Among the telescopes there are NASA's Infra Red Telescope Facility (IRTF), the United Kingdom Infrared Telescope (UKIRT), the Canada–France–Hawaii Telescope (CFHT), and the Japanese Subaru Telescope.

The first really successful orbiting infrared telescope was the US–Dutch–UK Infrared Astronomical Satellite (IRAS), launched in 1983 (see the illustration opposite). It had mirrors made of the metal beryllium rather than glass to cope with the very low temperatures (the infrared detectors were cooled in liquid helium down to a temperature of –269°C—–452°F). It operated for 10 months and scanned 96 percent of the sky before it ran out of coolant. The European Space Agency's Infrared Space Observatory (ISO) was put into orbit in 1995. Its main instrument was a camera, in addition to a telescope. The so-called cosmic microwave background radiation left over from the Big Bang at the creation of the universe has been mapped at long infrared and submillimeter wavelengths by NASA's Cosmic Background Explorer (COBE) satellite.

▼ **This thermogram** of someone eating ice cream records a range of temperatures from the warmest (red), through yellow and green, to the coldest (purple).

Electron microscopes

An electron microscope forms an image with a beam of electrons rather than the beam of light that passes through an ordinary optical microscope. And instead of glass lenses it has special magnets that focus the electrons to form an image, just as glass lenses focus light.

Optical microscopes (described in detail in Volume 5 of *Science Tools*) use glass lenses to focus beams of light to produce magnified images of small objects. But there is a limit to how much magnification can be achieved. This limit has to do with the fact that light is a wave motion, because the maximum possible magnification depends on the wavelength of light. Visible light has wavelengths between 380 nanometers (nm) at the violet end of the spectrum and 780 nm at the red end.

A beam of high-energy electrons also has a wavelength, but a much shorter one—typically, 0.004 nm. As a result, a microscope that uses an electron beam instead of a light beam can give extremely high magnifications. A beam of electrons carries a negative electric charge, so it can be deflected, or bent, by a magnetic field. In

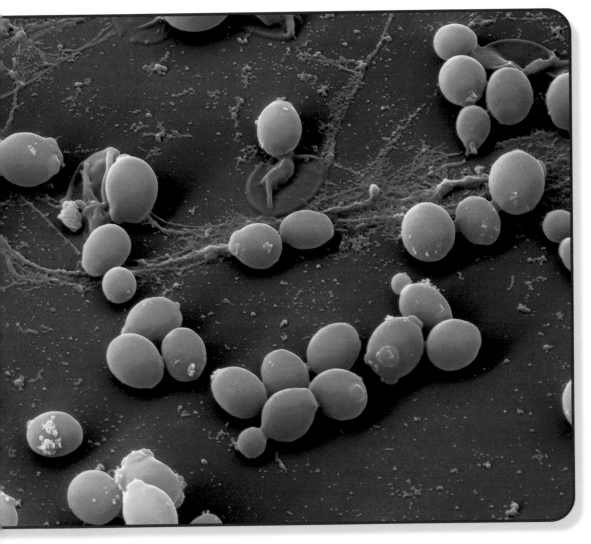

◄ **A microscopic fungus** called *Candida* is often found in moist parts of the human body. Its numbers are normally kept in check by the action of bacteria. But if the fungus multiplies, it can cause the disorder moniliasis (also called thrush). It is shown in this scanning electron micrograph, magnified 7,000 times.

place of the glass lenses of the optical microscope an electron microscope has magnetic "lenses" to focus the electron beam and produce an image. The first practical electron microscopes were built in the 1930s.

In most optical microscopes the light passes through the specimen, which therefore has to be transparent. (Most materials are transparent to visible light if they are sliced very thinly, so specimens for examination in a microscope are usually thin "sections.") In a transmission electron microscope (TEM) the electron beam also passes through the specimen—the specimen *transmits* the electrons through it. In another type, called a scanning electron microscope (SEM), the surface of the specimen is *scanned* by an electron beam. Electrons that bounce off its surface are collected and processed by a computer to produce an image of the specimen.

Transmission electron microscope

The illustration on this page shows the main parts of a transmission electron microscope. A vacuum pump extracts all the air from inside the instrument. An electron gun at the top of the microscope generates a stream of electrons, and magnetic lenses focus the electrons into a parallel beam. These magnetic lenses do not move for focusing. They consist of electromagnets, and the strength of the lens is altered by changing the electric current flowing in the electromagnets. The width of the beam depends on the size of the hole in the aperture plate through which it passes.

When the electrons strike the specimen (which has to be extremely thin), some of them bounce off, but most are transmitted through it and are focused by a second lens system. These electrons form an image on a detector or on a phosphorescent screen that can be viewed

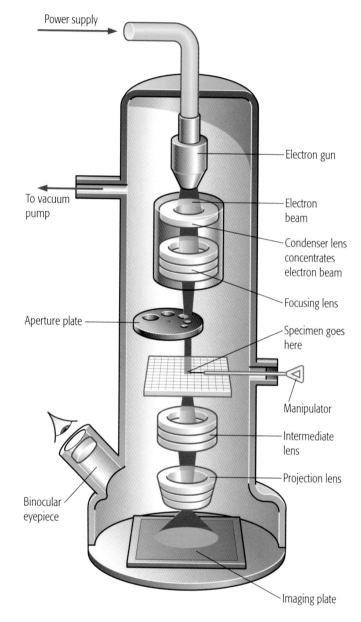

Power supply

Electron gun

To vacuum pump

Electron beam

Condenser lens concentrates electron beam

Focusing lens

Aperture plate

Specimen goes here

Manipulator

Intermediate lens

Projection lens

Binocular eyepiece

Imaging plate

▲ **In a transmission electron microscope** (TEM) a beam of electrons passes through a very thin slice of the object being studied.

FOR MORE ON ELECTRON MICROSCOPES SEE *FORCES BETWEEN ATOMS* **3**:*32*; *MEASURING CHARGE* **4**:*10*; *COMPOUND MICROSCOPE* **5**:*16*

Electron microscopes

through an eyepiece. The transmission electron microscope can magnify a specimen up to 1 million times (often written as ×1,000,000). Put another way, it can produce images of objects that measure less than 1 nanometer across.

Scanning electron microscope

The maximum magnification of a scanning electron microscope is about ×100,000. The top half of the instrument resembles a TEM (see the illustration on page 21). But after passing through the aperture plate, the electron beam is made to pass between two pairs of scan coils. They make the beam scan across the specimen in much the same way as the electron beam in a television set scans across the screen to produce an image.

The electrons scan parallel strips across the surface of the specimen, which does not have to be a thin slice. Some electrons bounce off the specimen, while others cause so-called secondary

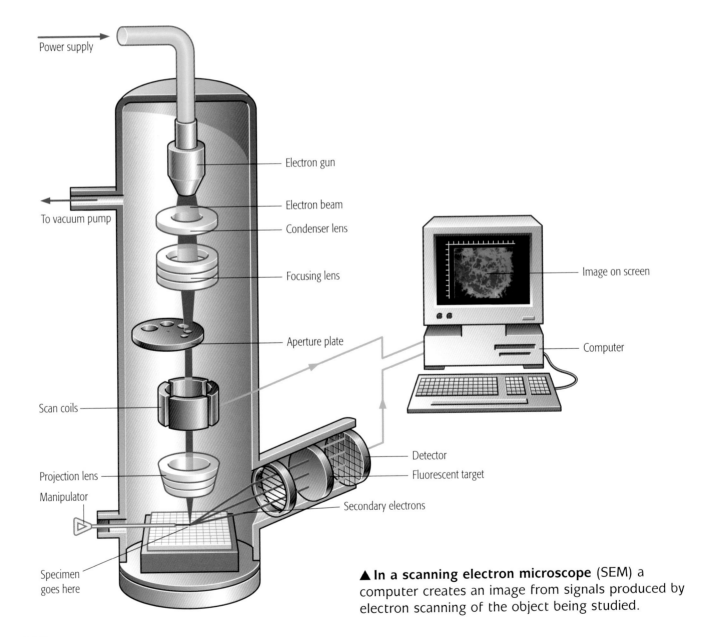

Power supply

To vacuum pump

Electron gun

Electron beam

Condenser lens

Focusing lens

Aperture plate

Image on screen

Computer

Scan coils

Projection lens

Manipulator

Detector

Fluorescent target

Secondary electrons

Specimen goes here

▲ **In a scanning electron microscope** (SEM) a computer creates an image from signals produced by electron scanning of the object being studied.

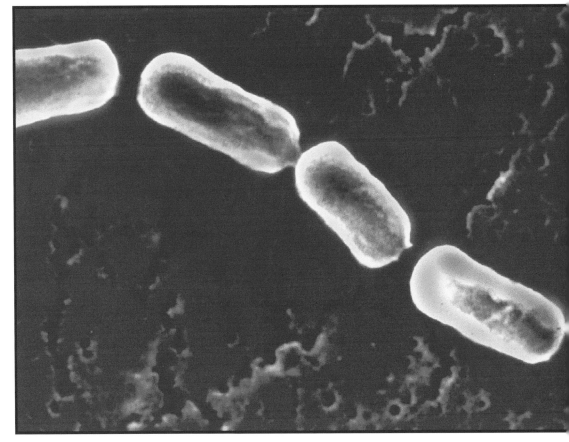

▶ A string of rod-shaped bacteria. This particular bacillus causes the life-threatening form of food poisoning known as botulism. In this scanning electron micrograph the bacteria are enlarged nearly 9,000 times.

electrons to be emitted from its surface. Both types of electrons are collected and counted by a detector, which sends a signal to a computer. Each point on the specimen produces a point of light on the computer's monitor, where a picture is gradually built up. As you can see from the pictures on these pages, the microscope can produce detailed three-dimensional images of the surfaces of objects down to about 10–20 nm in size. Specimens that conduct electricity give the best pictures. This is achieved by coating the specimen with a thin film of metal. The metal is evaporated onto the specimen in a vacuum chamber before the specimen is examined.

Scanning probe microscopes

In another type of scanning electron microscope a very sharp metal point called a probe scans close to the surface of the specimen. The point, usually made of the metal tungsten, is so sharp that its tip may consist of a single atom of metal. In the scanning tunneling microscope (STM), electrons "tunnel" between the probe and the sample, which must be able to conduct electricity. The resulting electrical signal is kept constant by raising and lowering the probe. In this way the probe is able to track the profile of the surface and, with the help of a computer, to produce a "contour map" of the surface. The STM is so sensitive that it can image individual atoms. Single atoms can also be imaged by a scanning transmission electron microscope (STEM), which can cope with very thick specimens. This instrument combines features of the scanning electron microscope and the transmission electron microscope.

Electron microscopes

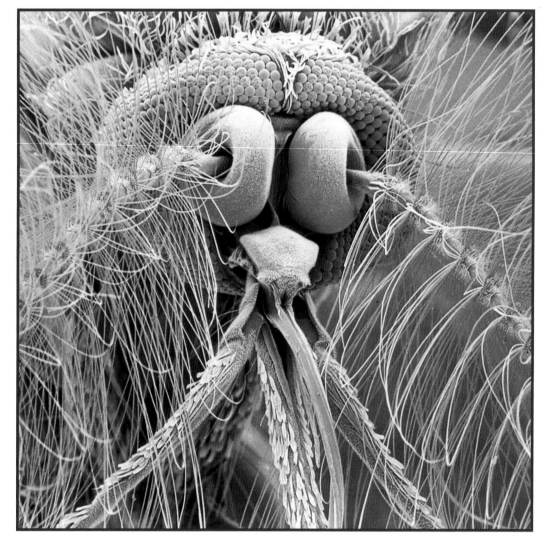

► **A hairy monster**—actually, a scanning electron micrograph of the head of a mosquito, magnified nearly 1,000 times. The blue facets near the top are its large compound eyes, while the feathery yellow structures are its sensitive antennae.

Nonmetallic specimens can be scanned using an atomic force microscope (AFM), which relies on mechanical forces instead of electrical signals. The probe is a tiny diamond chip, kept in contact with the specimen's surface by a lever and a spring. As the probe scans slowly across the surface, a computer monitors the force between the probe and the surface. The probe is continually raised or lowered so as to keep the force constant, and a computer converts these movements into a contour map of the surface.

When a specimen is bombarded with electrons, it gives off high-energy x-rays. The wavelength of the x-rays is a "fingerprint" of the atoms or molecules that produce them. In an electron probe microanalyzer, first developed in 1947, an x-ray spectrum analyzer is attached to an electron microscope. In this way the instrument not only provides a highly magnified image of a specimen, but also gathers information about its chemical makeup.

The inventors of electron microscopes

The first electron microscope was made in 1933 by the German engineer Ernst Ruska (1906–1988). He used a series of electromagnetic electron lenses, which he had developed two years earlier, and his microscope magnified about 7,000 times. At the

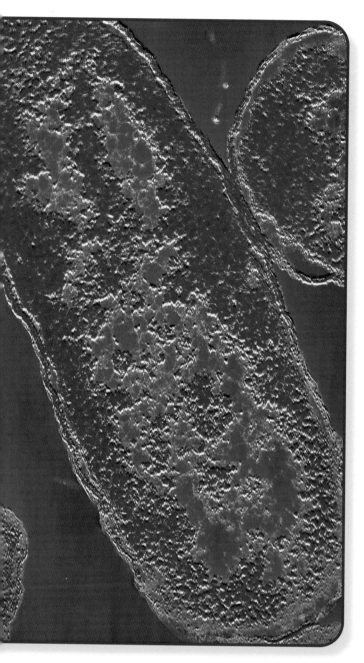

▲ **The bacterium *E. coli*,** commonly found in the intestines of humans. It is often used in experiments by microbiologists and genetic engineers. This transmission electron micrograph shows *E. coli* enlarged an amazing 100,000 times.

time, he was working at the Technical University of Berlin, but in 1937 he left to become a research engineer at Siemens AG, the company that in 1939 launched the first commercial transmission electron microscope. Ruska remained at Siemens until 1955, when he joined the Institute for Electron Microscopy of the Fritz Haber Institute as research director, where he remained until his retirement in 1972. In 1986 he shared the Nobel Prize for Physics, which was awarded for his work in electron microscopy.

The outbreak of World War II in Europe in 1939 brought Ruska's researches to an end. But development of the commercial electron microscope was continued in the U.S. by James Hillier. Hillier, a Canadian physicist born in 1915 in Brantford, Ontario, became a U.S. citizen in 1945. He worked for the Radio Corporation of America (RCA) from 1940 to 1953, and it was there that he constructed the first high-resolution electron microscope in the U.S. In 1941 responsibility for the production of an industrial version of the instrument passed to the Russian-born U.S. inventor Vladimir Zworykin (1889–1982), who had pioneered the development of electronic television for RCA in the 1930s. By 1946 Hillier had developed instruments with the maximum theoretical magnification possible.

One of Ruska's Nobel co-prizewinners in 1986 was Heinrich Rohrer, a physicist who was born in Switzerland in 1933, the year in which Ruska made his first microscope. Rohrer finished his studies at the Swiss Federal Institute of Technology, Zurich, in 1960, and in 1963 he joined the IBM Research Laboratory there. In 1978 he was joined by Gerd Binnig, a 31-year-old German physicist from Frankfurt. Together the two scientists constructed the first scanning tunneling microscope (STM), in 1981.

X-rays and gamma rays

X-rays and gamma rays have the shortest wavelengths in the electromagnetic spectrum. They can penetrate matter such as body tissues, as can be seen from x-ray photographs. They can also destroy tissue, so they are used in radiotherapy treatment of tumors and other abnormal growths.

At the end of the 19th century the German physicist Wilhelm Röntgen carried out experiments with cathode rays. This was the early name for beams of electrons given off by the cathode (negative electrode) of a discharge tube. He noticed that a fluorescent screen nearby began to glow with an eerie light and realized that some form of invisible radiation was causing the effect. Röntgen had accidentally discovered x-rays. At one time they were called roentgen rays in his honor.

X-rays have wavelengths from 10 nanometers (nm) to as little as 0.01 nm. The shorter the wavelength, the higher the energy. As a result, short-wavelength x-rays, called hard x-rays, have the greater penetrating power and pass through most kinds of matter (except heavy metals). The longer-wavelength soft x-rays are not so penetrating. Gamma rays have even shorter wavelengths than x-rays—between 0.1 and 0.00001 nanometers—and are extremely penetrating. They have up to 10 million times as much energy as a beam of visible light and can pass right through a sheet of lead 1 centimeter (0.4 in.) thick.

Production and uses of x-rays and gamma rays

X-rays are still produced by first creating a beam of electrons (see the illustration on the opposite page). The x-ray tube, invented in 1913 by the American physicist William Coolidge, contains a vacuum. An electron beam strikes the anode, called the target, which is made of the metal tungsten. As a result, x-rays are given off, and the tungsten target gets hot. In early x-ray tubes cold water was passed through channels inside the anode to keep it cool. In the modern tube an electric motor rotates the anode so that as each part is in turn struck by the electron beam, it has time to cool down before it next comes into the line of fire.

Wilhelm Konrad Röntgen

Wilhelm Röntgen (pronounced rurnt-gen*) was born in Lennep, Germany, in 1845. He studied engineering in the Netherlands and Switzerland before turning to physics and becoming a professor, finally holding a post at Munich University. He made his most famous discovery in 1895, when he noticed that invisible radiation from a high-voltage vacuum tube caused a fluorescent screen to glow. He called the new radiation x-rays—"x" for "unknown." For this discovery Röntgen was awarded the first Nobel Prize for Physics, in 1901, but he died a very poor man in 1923 during Germany's economic depression. The roentgen (symbol R), a former unit of radiation dose, was named in his honor.*

There are natural sources of x-rays in outer space. Certain celestial objects give off x-rays, for example, a pair of stars orbiting each other, one of which is a neutron star or a black hole. These and other x-ray sources are studied by x-ray astronomers using instruments on board orbiting satellites, because x-rays do not get through the Earth's atmosphere.

X-rays have many applications in research, industry, and medicine. Physicists use them to study the structures of crystals and metals. X-rays are used by chemists to identify elements and isotopes, and to detect fake gemstones. In industry they are used to detect flaws in metal castings and underground pipes. They can also be used to measure the thickness of materials and even to detect whether a sealed package of soap flakes is full or not. At airports x-rays allow officials to see inside the baggage of passengers they suspect might be carrying arms, explosives, or drugs. For all these applications x-rays are generated in x-ray tubes that work in essentially the same way as the one illustrated above right.

Gamma rays are produced naturally by some radioactive elements. In fact, it is the gamma rays in radioactivity that make it so dangerous. Some very high-energy gamma rays reach the Earth in cosmic rays from outer space. Gamma rays are used for sterilizing medical instruments and for killing germs in foods.

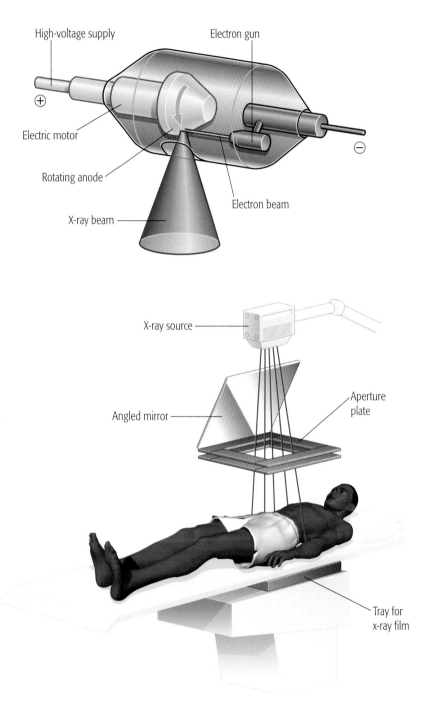

▲ **An electric motor** (top) rotates the anode of an x-ray tube to prevent it from getting too hot. An angled mirror (bottom) allows the radiographer to see the area of the patient's body that is being x-rayed.

FOR MORE ON X-RAYS AND GAMMA RAYS SEE *A LONG TIME AGO* **2**:*34*; *FORCES BETWEEN ATOMS* **3**:*32*; *FORCES INSIDE ATOMS* **3**:*36*

Detecting x-rays

A fluorescent screen, working on the same principle as the one first used by Röntgen, is still used to detect x-rays. So that a radiologist can check for lung disorders, a patient stands in front of a fluorescent screen that displays a "picture" of the chest and lungs. In a scintillation counter x-rays cause small flashes of light (scintillations) to appear on a fluorescent screen. The flashes can be counted electronically, giving scientists a way of measuring the intensity of an x-ray beam.

X-rays can also be detected by photographic film. They affect the film just as visible light does, causing it to "fog," or turn dark. X-ray photographs make use of this effect. Radiologists and other people who work with x-rays wear a film badge clipped to a lapel or pocket. When the film is developed, it indicates the person's level of exposure to x-rays.

▼ **The output from a CAT scanner** is a computer signal, so it can be sent over telephone lines just like an e-mail. Here a radiologist examines a scan transmitted from a hospital many miles away.

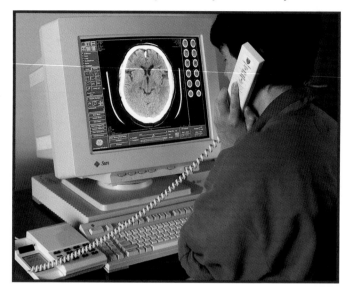

▶ **A CAT scan** is a painless procedure in which an x-ray source circles around the patient's body while he or she passes slowly through the machine on a movable table.

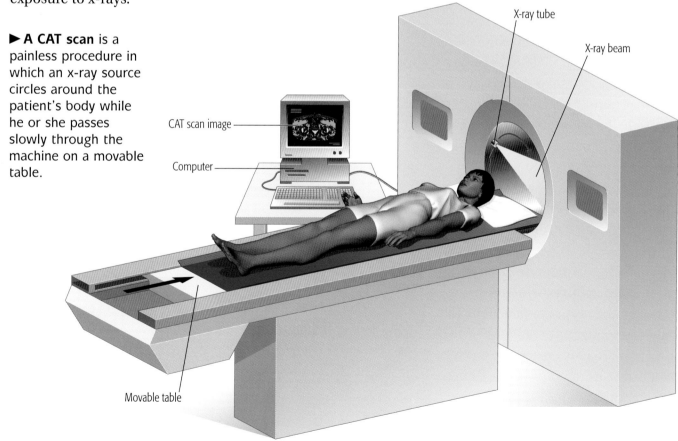

X-ray tube

X-ray beam

CAT scan image

Computer

Movable table

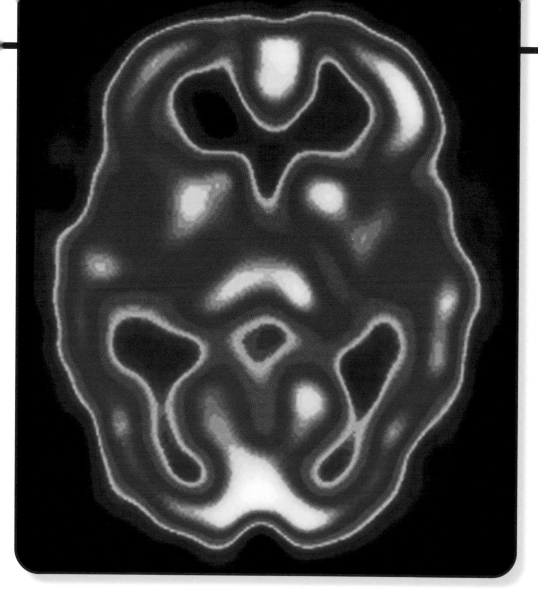

◀ **A SPECT scan** (easier to say than the full version, single photon emission computerized tomography) can reveal the activity of the living brain. Red and yellow areas indicate high brain activity, while gray and blue represent regions of low activity.

Medical uses of x-rays

There are two major medical uses of x-rays. They are employed in diagnosis—called radiology—and in treatment—called radiotherapy. The traditional type of x-ray machine used for diagnosis takes photographs of body structures, such as bones or internal organs. Bones tend to block x-rays, which pass more easily through softer tissues. On the x-ray film the areas where images of bones and hard tissue are formed do not get as much exposure as the areas of soft tissue and as a result appear bright against a dark background. (An x-ray photograph is a negative; there is no advantage in making a positive print from it.) Dentists also use x-rays to examine teeth.

Sometimes a doctor will introduce into a patient's body a substance that is opaque to x-rays. That is to make soft tissues show up. For example, a barium meal makes the stomach show up on an x-ray, and iodine compounds reveal the kidneys. To make an angiogram, a doctor injects a substance into an artery so that it travels to the patient's heart and reveals details of the coronary arteries on an x-ray.

Medical x-ray film is coated with photographic emulsion on both sides to make sure that a good image can be obtained with as little exposure as possible. That is because prolonged exposure, even to low-energy x-rays, can be harmful—which is why radiologists stand behind a protective screen when they are taking x-rays. Doctors ask for x-ray pictures only when they think that their usefulness in making a diagnosis far outweighs the very slight risk to the patient.

CATs and PETs

The use of x-rays to produce images of soft body structures is called tomography. The letters CAT are short for computerized axial tomography, the name of a technique in which a computer creates cross-sectional x-ray views through a patient's body. One or more x-ray tubes are mounted in a circular frame that rotates around the patient. Each rotation results in an image of a section, a "slice," through the body, as the patient is moved slowly through the machine. The computer assembles a set of slices to form a complete image, which can be three-dimensional and in color. The complete setup is illustrated on page 28.

A similar technique, based on a different type of radiation, is known as positron emission tomography, or PET for short. A radioactive tracer, such as radioactive glucose, is introduced into the patient's bloodstream from a drip. The glucose is made radioactive by incorporating some radiofluorine atoms. The glucose accumulates in the parts of the body to be imaged, such as the brain. There it gives off particles called positrons, which collide with any nearby electrons and emit very small amounts of gamma rays. Detectors in a circular frame surrounding the patient pick up these gamma rays. A computer analyzes the signals from the detectors and assembles a cross-sectional picture, rather like a CAT scan.

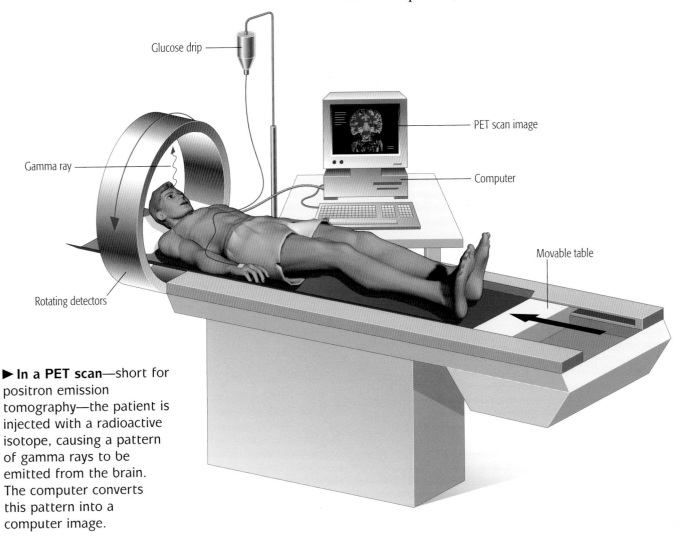

Glucose drip

PET scan image

Gamma ray

Computer

Movable table

Rotating detectors

▶ **In a PET scan**—short for positron emission tomography—the patient is injected with a radioactive isotope, causing a pattern of gamma rays to be emitted from the brain. The computer converts this pattern into a computer image.

◀**A computer screen** shows a radiologist successive "slices" through the brain of a patient undergoing a PET scan. The images can be stored on the computer and referred to later if necessary.

▼ **A CAT scanner** is a large x-ray machine that the patient is slowly moved through while the scan is being made.

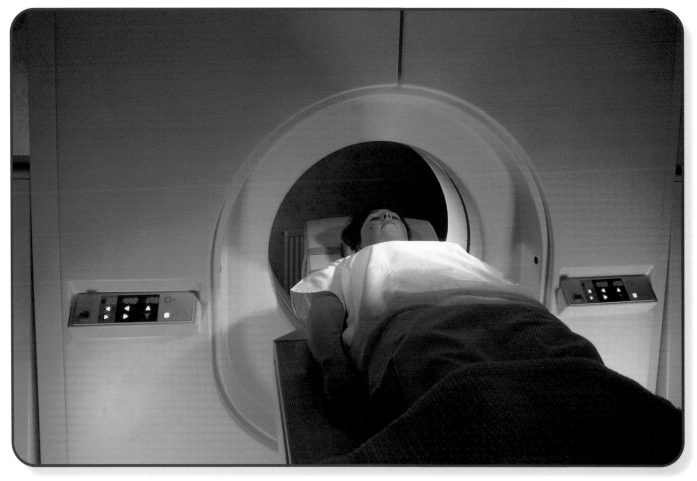

Using microwaves

Microwaves are radio waves of short wavelength, longer than for infrared radiation but shorter than for regular radio waves. They have a wide number of practical uses, from cellphones and satellite communications to radar equipment and microwave ovens.

Microwaves occupy the part of the electromagnetic spectrum between wavelengths of about 0.001 meters (a millimeter) and 0.03 meters (three centimeters). This is the short-wavelength end of the radio region. A special type of radio transmitter tube called a magnetron can be used to produce microwaves (see the illustration on this page).

Inside a magnetron a heated filament generates electrons. The electrons become concentrated into a sort of negatively charged cloud by the magnetic field between a pair of magnets above and below it. The field also makes the electron cloud move around the filament. Rapidly changing electric charges, produced as the electron cloud passes vanes in the anode block, generate an

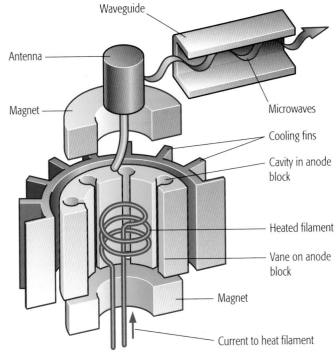

▲ **A type of vacuum tube** called a cavity magnetron is used to generate high-frequency microwaves.

◄ **Radar** is one of the commonest and most important uses of microwaves. It is essential for air traffic control and safety at airports.

electromagnetic field that oscillates at microwave frequencies within the cavities in the block. An antenna picks up the oscillations and gives off microwaves. Microwave radiation can be transmitted without losing its intensity by channeling it along metal tubes called waveguides.

Properties of microwaves

The chief property of microwaves is that they travel in straight lines. A radio engineer would say that they can travel only along lines of sight, which means that the transmitter and receiver must be within sight of each other. In practical terms, on land this limits their range to about 80 kilometers (50 miles). However, anything up in the sky is within sight, and for this reason microwaves are used for communicating with orbiting satellites and for radar signals. Also, ground links can be made longer by placing microwave antennas on tall masts.

Another unusual property of microwaves is the effect they have on water: Microwaves with a frequency of about 2,500 megahertz (MHz) set water molecules spinning. Friction between the rotating molecules generates heat; and if the water is in food, the food heats up and starts to cook. That is how a microwave oven works. It is also why the food gets hot but not the plate it is sitting on—the material from which the plate is made contains no water.

Microwave echoes

Radar (an abbreviation of "radio detection and ranging") locates moving objects, such as ships or airplanes, or even rain, by sending out pulses of microwave signals and detecting any echoes that bounce back. The system is similar in principle to sonar and the natural echo-location systems of some bats. A radar set uses the same antenna to

▲ **Engineers test** a microwave antenna that will be fitted to a satellite. The room has radiation-absorbing spikes on the walls, ceiling, and floor so that no signal is reflected from them to the antenna.

FOR MORE ON USING MICROWAVES SEE *RADIO TELESCOPES* **6**:36; *REPRODUCING SOUND* **7**:20; *ORGANIZING THE HEAVENS* **9**:34

send and receive. The antenna is usually shaped like a dish, and at its center is a transponder. It is a device that alternately sends and receives signals. The dish is usually steerable, so that it can be aimed at the required target. It may rotate to scan the sky continuously, or it may have electronic circuits that make the radar beam scan from a stationary antenna.

The cellphone network

Microwave links are essential for cellphone networks. The illustration below shows how they work. Each cell in the system has a base station with microwave transmitters and receivers. Suppose that someone on the red phone wants to call someone on the blue phone. When he or she

punches in the number, a message travels to the base station, which passes it on to the nearest cellphone exchange. Here the call is routed to the blue phone's base station, from where it is transmitted to the blue phone.

But what if the phone you are calling is not connected to your cellphone exchange, like the green phone on the right on page 35? Then your cellphone exchange passes the call to a main exchange. That main exchange is connected by a landline or microwave link to the main exchange near the green phone. The signal then passes to a cellphone exchange, which routes it via a base station to the green phone.

What if the green phone on page 35 moves to the left (perhaps it is being used in a car or on a

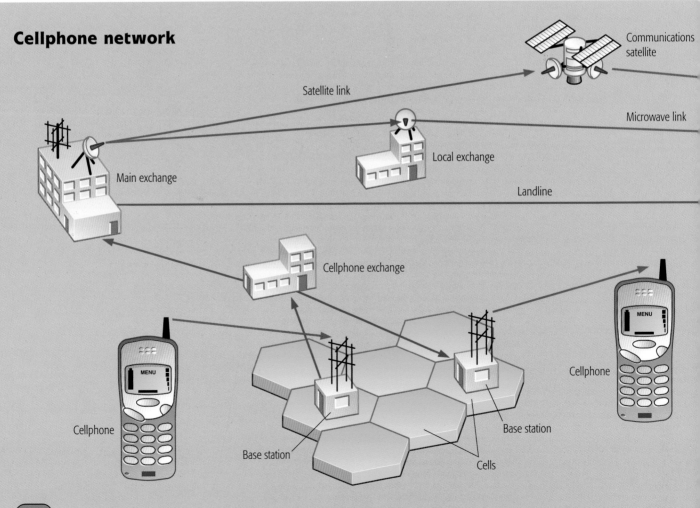

Cellphone network

Communications satellite

Satellite link

Microwave link

Local exchange

Main exchange

Landline

Cellphone exchange

Cellphone

MENU

Cellphone

MENU

Base station

Base station

Cells

Cellphone

train)? Once the signal from the moving phone falls below a certain strength, the call is automatically routed to a different base station so as to maintain good contact with the green phone.

And what happens if the two main exchanges are too far apart for a landline or microwave link—what if they are on different continents, say? The first main exchange beams the signal up to a communications satellite, which relays it to a main exchange near the green phone. And the whole process takes place automatically without any telephone operators!

▶ **A cellphone** allows people to talk to each other wherever they are in the world—almost!

A cellphone network uses a combination of landlines (underground cables) and microwave radio links.

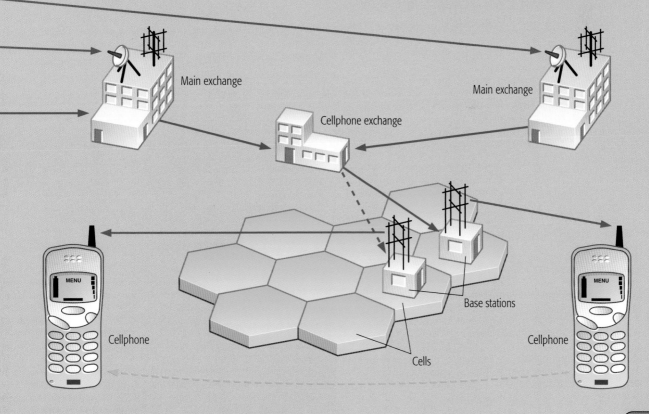

Radio telescopes

Optical telescopes are used to view celestial objects that give off visible light, such as stars, or reflect sunlight, such as the Moon. But many stars and galaxies emit invisible radiation at the radio end of the electromagnetic spectrum and can be "viewed" only by using radio telescopes.

Just over a hundred years ago scientists discovered how to produce radio waves—a type of electromagnetic radiation with the longest wavelengths of all, from tens of centimeters to 10,000 meters. Since that time radio and its further development, television, have become the most important communication medium on Earth and the source of most of the radio waves that permeate the air around us. But there are other, extraterrestrial sources of radio waves, and they are studied by radio astronomers.

Radio astronomy was born in 1932, when an American engineer, Karl Jansky, figured out that some of the radio "noise" picked up by the antenna he was using came from the Milky Way. Unlike many forms of electromagnetic radiation, radio waves pass straight through the Earth's atmosphere and reach the ground. Typical radio wavelengths from space are about 10 meters (33 feet). The best type of antenna for receiving them is dish-shaped, and the dish's diameter has to be as big as the wavelength. Since the late 1950s, large radio telescope dishes have been constructed at various places around the world.

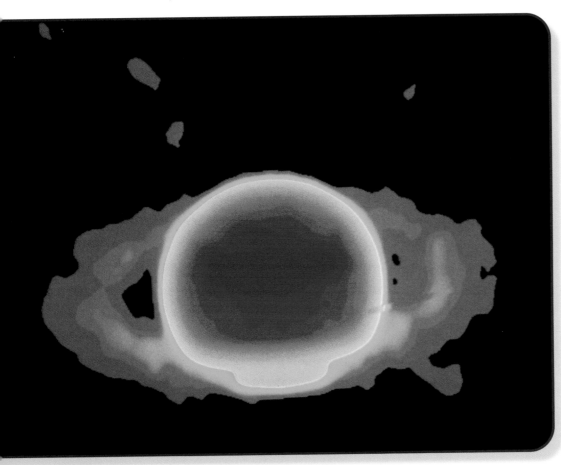

◄ **This radio telescope image** of Saturn shows, in blue, the rings of fine particles that encircle the planet. This is a so-called false-color image, with colors added by a computer to show how the strength of the radio signal varies across the planet.

► **A radio telescope** consists essentially of a large dish-shaped antenna that can be aimed at any point in the sky to detect radio signals from space.

◄**The Very Large Array** (VLA) at Socorro, New Mexico, is the world's most complex radio telescope on one site. It has 27 separate dishes, each 25 meters (82 feet) across.

FOR MORE ON RADIO TELESCOPES SEE *REFLECTING TELESCOPES* **5**:*28*; *USING MICROWAVES* **6**:*32*; *ORGANIZING THE HEAVENS* **9**:*34*

Radio telescopes

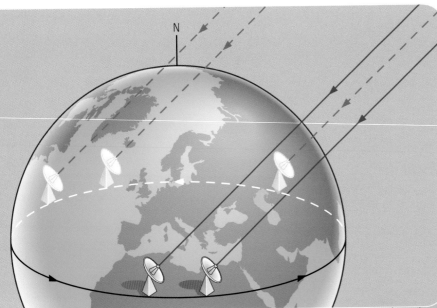

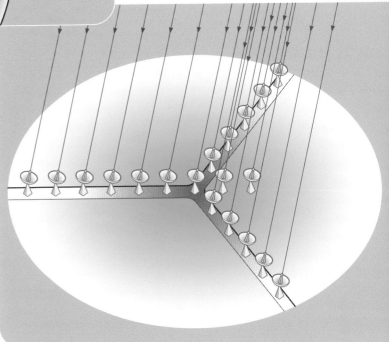

◀ **Two radio telescopes** can be used as an interferometer by taking readings from different places as the rotation of the Earth carries them around.

▼ **A large number of dishes** on a Y-shaped railroad track—the layout of the Very Large Array. Each arm of the "Y" is about 20 kilometers (12.4 miles) long. The dishes can be used together to gain the same results as from a single dish 27 kilometers (nearly 17 miles) across.

The largest, at 305 meters (1,000 ft) across, is located at Arecibo, Puerto Rico, in an old volcano crater.

Sources of radio waves

There are many sources of radio waves from space. The Sun and some planets emit radio waves. Within our own galaxy (the Milky Way) clouds of gas left over from the explosion of supernovas generate strong radio signals. So do pulsars, which are rapidly spinning stars that give off radio waves in a beam that rotates like the light from a lighthouse, so that they arrive as a series of pulses. Sources beyond the Milky Way include radio galaxies and quasars. Quasars (the name is short for "quasi-stellar objects") look like stars but are in fact the bright centers of immensely distant galaxies at the outer fringes of the Universe.

Types of telescope

The commonest type of radio telescope is a single steerable dish. Large electric motors aim the dish

at the correct point in the sky. The dish focuses radio waves just as a concave mirror focuses light. The receiver that picks up the radio signal is at the focus of the dish, mounted on struts. In an alternate arrangement a subreflector at the focus bounces the radio waves down to a receiver at the center of the dish. The radio signal passes from the receiver to the control room, which may be located below the telescope or in another building

▲ **The largest single radio telescope** in the world is built into a volcano crater at Arecibo in Puerto Rico. The receiving apparatus is supported over the dish on wires suspended from three tall towers.

nearby. There it is amplified and converted into a digital signal that is processed by a computer to produce an image.

The sheer weight of the structure that has to be supported limits the size of single steerable dishes. However, a very large dish can be mimicked by combining several smaller dishes to make a so-called radio interferometer. When two dishes are aimed at the same object, for example, the two radio signals can be added together. One method uses the rotation of the Earth to move the dishes to different positions for collecting multiple signals. In another, many small dishes spread out in a large array can be made to operate as if they were one huge radio telescope. The technique, which is called aperture synthesis, needs a very powerful computer to handle all the signals and combine them to produce a single image. Using this method, radio astronomers can "see" objects just one-hundredth the size of those visible with the world's best optical telescopes.

▲ **A close-up of the receiver** of the Arecibo telescope shows the new reflector system installed in 1997. It is raised above the fixed dish when in use.

Glossary

Any of the words in SMALL CAPITAL LETTERS can be looked up in this Glossary.

aperture synthesis The use of several small RADIO TELESCOPES to produce the same effect as using one very large radio telescope.

bolometer A sensitive instrument that measures radiant heat, such as INFRA-RED RADIATION.

CAT scanner A medical X-RAY scanner that produces cross-sectional images of the body's soft tissues, such as the heart and lungs. CAT stands for **c**omputerized **a**xial **t**omography.

cathode rays Streams of ELECTRONS produced by the cathode of a vacuum tube such as an X-RAY TUBE or cathode-ray tube.

cellphone A mobile tele-phone that makes use of MICROWAVE radio transmitters and receivers located in "cells" that cover a region.

CFC Short for **c**hlorofluoro-**c**arbon, an organic chemical used as a propellant gas in aerosols and thought to be a cause of damage to the OZONE LAYER.

cosmic rays Radiation from outer space that consists mainly of high-energy particles with some GAMMA RAYS.

electromagnetic radiation Any radiation that travels as WAVES, ranging from short-wavelength GAMMA RAYS to long-wavelength RADIO WAVES. It includes visible light, as well as ULTRAVIOLET RADIATION and INFRARED RADIATION.

electromagnetic spectrum The whole range of FREQUENCIES across which ELECTROMAGNETIC RADIATION is found.

electron A SUBATOMIC PARTICLE that has a negative charge; it is the basic unit of electricity. Electrons surround the NUCLEUS of an atom.

electron gun A device for producing a stream of ELECTRONS in, for example, a cathode-ray tube, X-RAY TUBE, or ELECTRON MICROSCOPE.

electron microscope A type of microscope that uses a stream of ELECTRONS (instead of visible light) and electro-magnets (instead of glass lenses) to produce highly magnified images.

fluorescent screen A sheet of glass coated with a PHOSPHOR that gives off light when struck by streams of electrons or X-RAYS.

fluorescent tube A type of electric bulb in which high-energy atoms ("excited" by collisions with ELECTRONS) give off light that, in turn, makes a PHOSPHOR produce much more light. The final output can be visible light or ULTRAVIOLET RADIATION.

frequency For any wave motion the number of complete cycles of the wave that pass a given point every second. It is expressed in HERTZ.

gamma rays High-energy, short-wavelength, penetrat-ing ELECTROMAGNETIC RADIATION produced by some radioactive elements.

global warming A very gradual increase in the average temperature of the Earth and its atmosphere resulting from the GREENHOUSE EFFECT.

greenhouse effect An effect in the Earth's atmosphere in which certain gases (such as carbon dioxide from atmospheric pollution) absorb INFRARED RADIATION emitted by the Earth's surface. This traps heat and results in GLOBAL WARMING.

heat rays Another name for the INFRARED RADIATION given off by all hot objects.

hertz (symbol **Hz**) The SI unit of FREQUENCY. One hertz equals a frequency of one cycle of the wave or oscillation per second.

image intensifier An electronic instrument in which an image illuminated with very dim light or invisible light (such as INFRARED RADIATION) is converted into a visible image on a FLUORESCENT SCREEN.

infrared astronomy The study of celestial objects that emit INFRARED RADIATION, often using space probes and satellites.

infrared radiation A type of ELECTROMAGNETIC RADIATION with wavelengths just longer than those of visible light, as given off by all hot objects.

landline A telecommuni-cations link in the form of a wire (cable) or fiber-optic cable.

magnetron A type of vacu-um tube used to generate high-power MICROWAVES (mainly for use in radar).

microwave background radiation A weak MICROWAVE radiation from the universe, thought to be left over from the Big Bang, in which the universe was formed.

microwaves A type of ELECTROMAGNETIC RADIATION in the form of very short-wavelength RADIO WAVES.

nucleus The positively charged central part of an atom, made up of PROTONS and NEUTRONS (except in hydrogen, whose nucleus contains a single proton).

ozone layer A layer in the Earth's atmosphere at an altitude of about 15–50 kilometers (roughly 10–30 miles) that contains most of the atmosphere's ozone gas. It blocks out most ULTRA-VIOLET RADIATION but is in danger of being damaged by CFCS.

PET scanner A type of medical scanner that pro-duces cross-sectional images of the body's soft tissues, such as the brain. The initials stand for **p**ositron **e**mission **t**omography.

phosphor A substance that gives off light when struck by streams of ELECTRONS or X-RAYS.

pulsar A celestial object that emits very regular pulses of RADIO WAVES. Pulsars are believed to be rapidly rotating neutron stars.

quasar A compact starlike object that produces intense ELECTROMAGNETIC RADIATION over a wide range of wave-lengths (from X-RAYS to RADIO WAVES).

radar An electronic device that detects the range and bearing (direction) of objects, such as airplanes in flight, which reflect back echoes of the MICROWAVE signals the radar set transmits.

radio astronomy The study of celestial objects that emit RADIO WAVES, generally using a RADIO TELESCOPE.

radio interferometer A high-resolution astronomical instrument consisting of two or more RADIO TELESCOPES aimed at the same radio source.

radio telescope An astronomical instrument that collects and records RADIO WAVES coming from space. The commonest type consists of a large parabolic reflector in the shape of a shallow dish.

radio waves A type of ELECTROMAGNETIC RADIATION with wavelengths longer than those of MICROWAVES.

radioactivity The emission of particles or radiation by atomic nuclei, which (except with gamma radiation) change into other kinds of nuclei as a result.

radiography The medical use of X-RAYS to take photographs of body structures.

radio-opaque Describing a substance that X-RAYS will not pass through, sometimes introduced into body organs to make them show up on an x-ray photograph.

radiotherapy The medical use of X-RAYS to treat certain disorders.

scanning electron microscope (SEM) A type of ELECTRON MICROSCOPE in which an electron beam scans across the surface of a specimen, which gives out "secondary electrons" that are converted into an image by a computer.

scintillation counter A device in which X-RAYS (or other radiation) cause tiny flashes of light to appear on a FLUORESCENT SCREEN. The flashes can be counted electronically to give a measure of the x-ray intensity.

spectrum A range of colors (light of different wavelengths) produced when white light is split by a prism or diffraction grating. See also ELECTROMAGNETIC SPECTRUM.

speed of light The speed of light in a vacuum, equal to nearly 300 million meters per second (nearly a billion feet per second). Symbol c.

subatomic particle Any of the particles that make up atoms, including ELECTRONS, NEUTRONS, and PROTONS.

supernova The sudden explosion of a star that completely destroys it, giving out huge quantities of light and other forms of ELECTROMAGNETIC RADIATION.

thermogram A photograph that records the INFRARED RADIATION given off by a hot object.

transmission electron microscope (TEM) A type of ELECTRON MICROSCOPE in which a beam of electrons passes through a very thin specimen and records a

highly magnified image of it on a photographic plate.

transponder An electronic device that can both transmit and receive signals.

ultraviolet astronomy The study of celestial objects that emit ULTRAVIOLET RADIATION, generally using space probes and satellites.

ultraviolet radiation A type of ELECTROMAGNETIC RADIATION with wavelengths just shorter than those of visible light.

visible radiation A type of ELECTROMAGNETIC RADIATION that humans can see; ordinary light.

wave A regular disturbance in space, or within a medium, that transfers energy in its direction of travel. For example, all types of ELECTROMAGNETIC RADIATION travel as waves.

waveguide A metal tube of circular or rectangular cross-section used to carry MICROWAVES (which will not pass along ordinary wires or cables).

wavelength The distance between two neighboring peaks (or troughs) of a traveling WAVE.

x-ray tube A type of vacuum tube used to produce X-RAYS.

x-rays A type of penetrating ELECTROMAGNETIC RADIATION with wavelengths slightly longer than those of GAMMA RAYS but shorter than those of ULTRAVIOLET RADIATION.

Set Index

Set Index

Set Index

Further reading/websites and picture credits

Further Reading

Atoms and Molecules by Philip Roxbee-Cox; E D C Publications, 1992.

Electricity and Magnetism (Smart Science) by Robert Sneddon; Heinemann, 1999.

Electronic Communication (Hello Out There) by Chris Oxlade; Franklin Watts, 1998.

Energy (Science Concepts) by Alvin Silverstein et al.; Twenty First Century, 1998.

A Handbook to the Universe: Explanations of Matter, Energy, Space, and Time for Beginning Scientific Thinkers by Richard Paul; Chicago Review Press, 1993.

Heat (How Things Work Series) by Andrew Dunn; Thomson Learning, 1992.

How Things Work: The Physics of Everyday Life by Louis A. Bloomfield; John Wiley & Sons, 2001.

Introduction to Light: The Physics of Light, Vision and Color by Gary Waldman; Dover Publications, 2002.

Light and Optics (Science) by Allan B. Cobb; Rosen Publishing Group, 2000.

Electricity and Magnetism (Fascinating Science Projects) by Bobbi Searle; Copper Beech Books, 2002.

Basic Physics: A Self-Teaching Guide by Karl F. Kuhm; John Wiley & Sons, 1996.

Eyewitness Visual Dictionaries: Physics by Jack Challoner; DK Publishing, 1995.

Makers of Science by Michael Allaby and Derek Gjertsen; Oxford University Press, 2002.

Physics Matters by John O.E. Clark et al.; Grolier Educational, 2001.

Science and Technology by Lisa Watts; E D C/Usborne, 1995.

Sound (Make It Work! Science) by Wendy Baker, John Barnes (Illustrator); Two-Can Publishing, 2000.

Websites

Astronomy questions and answers —
http://www.allexperts.com/getExpert.asp?Category=1360

Blood classification —
http://sln.fi.edu/biosci/blood/types.html

Chemical elements —
http://www.chemicalelements.com

Using and handling data —
http://www.mathsisfun.com/data.html

Diffraction grating —
http://hyperphysics.phy-astr.gsu.edu/hbase/phyopt/grating.html

How things work —
http://rabi.phys.virginia.edu/HTW/

Pressure —
http://ldaps.ivv.nasa.gov/Physics/pressure.html

About rainbows —
http://unidata.ucar.edu/staff/blynds/rnbow.html

Story of the Richter Scale —
http://www.dkonline.com/science/private/earthquest/contents/hall2.html

The rock cycle —
http://www.schoolchem.com/rk1.htm

Views of the solar system —
http://www.solarviews.com/eng/homepage.htm

The physics of sound —
http://www.glenbrook.k12.il.us/gbssci/phys/Class/sound/u11l2c.html

A definition of mass spectrometry —
http://www.sciex.com/products/about mass.htm

Walk through time. The evolution of time measurement —
http://physics.nist.gov/GenInt/Time/time.html

How does ultrasound work? —
http://www.imaginiscorp.com/ultrasound/index.asp?mode=1

X-ray astronomy —
http://www.xray.mpe.mpg.de/

Picture Credits

Abbreviation: SPL Science Photo Library